GOD

FOR THE CURIOUS UNBELIEVER

PAUL GEORGIOU

Published by Panarc International 2016
Copyright Paul Georgiou

First Edition

The author asserts the moral right under the Copyright, Designs and Patents Act 1988 to be identified as the author of this work.

All rights reserved. No part of this publication may be reproduced, stored in a retrieval system or transmitted in any form or by any means without the prior consent of the author, nor be otherwise circulated in any form of binding or cover other than that in which it is published and without similar condition being imposed on the subsequent purchaser.

Panarc Publishing
www.panarcpublishing.com

Panarc International
www.panarc.com

ISBN: 978-0-9954637-2-1

.epub eBook ISBN: 978-0-9954637-3-8
.mobi eBook ISBN: 978-0-9954637-4-5

GOD

FOR THE CURIOUS UNBELIEVER

PAUL GEORGIOU

CONTENTS

1. INTRODUCTION — 1

2. CRITERIA — 4
- 2.1 Facts — 4
- 2.2 Reason — 5
- 2.3 Natural v. supernatural — 5
- 2.4 Probability — 6
- 2.5 Summary — 7

3. KEY QUESTIONS — 9
- 3.1 Why is there something rather than nothing? — 9
  - 3.1.1 Theism — 12
  - 3.1.2 Deism — 13
  - 3.1.3 Atheism — 13
  - 3.1.4 Conclusion — 16
- 3.2 Why is there this particular something? — 17
  - 3.2.1 The universe as a 'put up job' — 23
  - 3.2.2 It would have to be the way it is, wouldn't it? — 24
  - 3.2.3 The universe with a sense of direction — 26
  - 3.2.4 The hindsight fallacy — 28
  - 3.2.5 Conclusion — 31
- 3.3 What exists? — 34
- 3.4 What are rules? — 38

4. HUMAN EXPERIENCE — 41
- 4.1 Half angel, half beast — 43
- 4.2 The arts — 45
  - 4.2.1 Emergent properties — 48
  - 4.2.2 Self-less consciousness — 49
- 4.3 Love — 51
- 4.4 Morality — 53
- 4.5 Conclusion — 57

5. GOD? 60
5.1 Should we take the concept of God seriously? 60
5.2 What could we know of God? 64
5.3 Questions about the nature of God 66
5.3.1 Creative and Powerful? 66
5.3.2 Benign? 67
5.3.3 Omnipotent? 69
5.3.4 Omniscient? 70
5.3.5 The Evolution of God 72
5.4 Conclusion 77

6. FINAL CONCLUSIONS 79

7. AN AFTERTHOUGHT 91

8. APPENDIX: OBJECTIONS & ANSWERS 94

1. INTRODUCTION

I'm not any good at faith.

It's not that I'm afraid to take a leap. It's just that, if you abandon evidence and reason, there's no way of knowing whether your leap is taking you in the right direction.

So that's it. I intend to take on the toughest questions (about the nature of existence) but, in answering them, I'm going to look at the facts and apply reason.

I'm inviting you to join me on a journey. What do you need? An open mind. A willingness to accept conclusions if they are clearly based on facts and reason, even if the conclusions are counter-intuitive.

There will be nothing complicated as we move forward. It's a journey anyone can take. Only those whose minds are closed will fall by the wayside.

Anyone who is hoping we will find absolute truth, whatever that means, will be disappointed. As we progress, we will have to make compromises at every stage. But the journey is still worthwhile because, at the end, although we won't have absolute truth, we will have come as close to it as is possible.

And we will have avoided the irrational excesses of the extremists on either side of the existential debate.

As you go through the sections of this book, there will be points at which you may feel my argument is defective, and you may be right, but in the Appendix, I attempt to answer some of the more frequent objections. There are number references to the answers to common objections at the end of any section dealing with particularly contentious issues.

(Objections: 1, 2, 3)

Warning: You will be asked to rethink many of the certainties which you take for granted and believe are true – because they are not true. The material world is not what you think it is. Space and time are not what you think they are. Simplistic answers to existential questions do not survive even cursory scrutiny of the facts. If you are determined to cling to these 'certainties', of course you may; but two consequences ensue:

Persistent belief in these provably false certainties is an act of faith even more irrational than a belief in God or fairies. (After all, we can disprove these spurious certainties; whereas we can't actually disprove the existence of God or fairies.)

This book is not for you.

But let's not get ahead of ourselves. For now at least, stay with me. Let's go.

2. CRITERIA

Before we start, we need to say a bit more about the ground rules.

2.1 Facts

By facts, we mean all that body of information which is considered to be true, either because there is proof something actually happened or because, using the scientific method, we can verify that a fact is consistent with reality.

Here is our first compromise. All facts are true *to the best of our knowledge*. Given how many 'facts' have subsequently been proved wrong, only a fool would be absolutist. But, in looking for answers to our questions, the best basis must be **the body of knowledge that we currently think is true.**

Some would argue that, since some 'facts' have turned out to be false, we can treat all facts with equal scepticism. This attitude fails the test of reason. Assuming we wish to understand our world and our place in it, we have to start with that body of factual knowledge which is currently thought to be true. The alternative would be to declare the quest for understanding futile.

(Objections: 3, 4, 6)

2.2 Reason

By reason, I mean the capacity to draw logical conclusions from clearly defined premises and to reject inconsistency.

Again, we have to recognise a limitation. Reason doesn't exist in a vacuum. We need a basis of fact in order to exercise reason. If we are to prove Socrates is mortal, we first need to 'know' both that a) 'All men are mortal.' and that b) 'Socrates was a man'. Only then, if we accept these factual premises, can we prove that Socrates was mortal.

So, like facts, reason is a powerful and essential tool in seeking answers to questions but we have to be aware of its limitations.

(Of course the Athenians in 399 BC didn't bother with logic. They gave Socrates hemlock and he died, thus establishing his mortality as a matter of fact beyond doubt.)

(Objection: 5)

2.3 Natural v. supernatural

It follows from our reliance on fact and reason that, whenever we are presented with a natural and a supernatural explanation, we will favour the natural. Why? Because, in general, the natural is based on observable evidence and is verifiable; whereas the supernatural is inherently speculative and unprovable.

2.4 Probability

This may be the most difficult criterion for some readers to accept but it is thoroughly endorsed by the facts and by reason.

In seeking answers, we will often have to settle for the most probable answer, i.e. if, after taking into account all the facts and rigorously applying reason, the truth still remains uncertain. In such a case, we will opt for the answer which, given the facts and applying reason, is most probable.

It's important you accept the probability criterion. There was a time, not so long ago, when we thought the universe was a giant machine, governed by the immutable and precise laws of cause and effect. We now know better.

At the most fundamental level, there is inherent uncertainty. As Heisenberg explained, we can know the momentum of a particle but not its precise position; and we can know its precise position but not its momentum. But we cannot know both at the same time. At the level of fundamental particles, we have to work with probabilities.

Many systems are so complex that it is impossible to reduce them to simple cause/effect relationships. Today, we have to settle for probabilities. There is a 70% chance of rain tomorrow. Each of us has a 50% chance of developing some form of cancer in the course of our lives. The chances of winning the lottery are 45,000,000 to one.

So, if we accord pre-eminence to the facts and reason, it is reasonable, when the facts and reason fail to determine which of various possible explanations we should choose, to include probability in our criteria.

There will be those who refuse to accept probability and I fully understand their position. They demand the certainty provided by facts and reason. They fear that any concession to probability will open up a Pandora's box of irrational speculation. I respect their position but I would say to them, read on. They will almost certainly find that the facts on which they predicate their position are not as they believe them to be. Or, if they are, they will have to admit that, despite their strict adherence to the facts and reason, their world view is already heavily reliant on probability.

(Objection: 7)

2.5 Summary

I want to strike a deal with you, the reader.

The method we will use to explore the fundamental questions about human existence will be:

- facts
- reason
- probability

We will not start with any preconceived notions. For example, we won't begin by insisting there is a God; nor will we insist there is no God. But we will agree that there are questions to be answered; and that only a lazy or blinkered mind can look at the universe and simply take it, and everything in it, for granted.

To address these fundamental questions, we will start by taking into account what we know to be true and then, by applying reason to it, we will then opt for the most probable explanation.

Whenever we are presented with a natural and a supernatural explanation, we will favour the natural.

At the end, we won't know 'the truth'. But we will know that our answers are more likely to be true than any other.

If you agree with these ground rules, we will set out on an extraordinary journey which will lead us in surprising and unexpected directions.

If you don't agree, go to Appendix 1 and, before you quit, see whether you find any of the answers to your objections persuasive.

3. KEY QUESTIONS

We begin our journey by addressing some key questions about the nature of existence and our perceptions of reality.

3.1 Why is there something rather than nothing?

First, let's agree there is something. It's probably one of the few things we can be absolutely certain of. It's an absolute fact. There is something. Even if everything we perceive is illusory, even if we ourselves are illusory, even if the whole of existence is simply a computer simulation, even so there is still something.

What precisely is this something? You might think we know the answer, namely a predominantly, if not exclusively, material universe. But science would hesitate to claim it knows the precise nature of material existence. The best answer science can give is not the stone that Dr Johnson kicked to demonstrate the existence of material things; it is a bizarre world of obscure particles behaving according to the rules of probability, rather than certainty. (More on this later.)

Not to worry. At least we are certain there is something.

So let's ask again. Why is there something rather than

nothing? It's one of the most searching and challenging questions we can ask.

Just think about it. If there was nothing, what would be the chances of there being something? Reason suggests the chances would be zero. After all, if there was nothing, there would be nothing to cause there to be something.

But there is something. That's a fact. So logically we must increase the odds of there being something from zero to at least infinitesimal.

On the other hand, some would argue that there was always something and therefore the odds of there being something are 100%. But, intellectually, that doesn't help. Given the alternative possibility of nothing, the fact that there is something remains extraordinarily improbable. The odds against winning the lottery do not change just because somebody has won it. So the odds of something coming out of nothing remain rationally zero and factually infinitesimal, even though we know it has happened.

And of course, the Big Bang theory, which postulates that the entire universe, including space and time, or space/time, came into being at a particular point in what we now call time (13.5 billion years ago), militates against the 'it's always been there, so I don't need to consider, much less answer, the question' argument.

Assuming the universe had a beginning, we are justified in also assuming something caused it to begin. It didn't exist; then it did exist. So something must have caused it. Those who prefer to say 'It just happened. It wasn't caused' are entitled to their view but it fails the tests of fact (the Big Bang occurred), reason (we expect an effect to have a cause) and probability (the only alternative, i.e. it just happened, is less probable than the theory of a cause).

What can we say of the coming into being of the universe? Not much, but more than nothing. We can say:

- whatever caused the universe was 'outside' time and space (because time/space came into being with the rest of the universe)
- it was immensely powerful (it generated from a single point – a singularity – a space/time continuum at least 13.5 billion light years across)
- it was inherently ordered (it performed according to discoverable laws of physics)
- it included an extraordinary creative capacity (after all, so far it has produced stars, planets, life and human consciousness)

Please note that none of the above depends on faith. It is based on facts, reason and probability. Anyone can put forward an alternative hypothesis but it will be easily exposed as:

- ignorant of the facts
- defective in reasoning
- inherently improbable

Let's take three obvious examples:

Theism: an omnipotent, omniscient and benign God created the world and all that is in it and currently maintains it

Deism: as for theism but, having set the universe in motion, God tends not to intervene

Atheism: the universe is just there, it is given. All we can do is work within it to understand it

3.1.1 Theism

Given the starting point for this book, the obvious objection to the theist's position is that, so far at least, there is no proof whatsoever of the existence of an omnipotent, omniscient and benign God.

It's true we have agreed that the most likely explanation for the universe (i.e. one based on the facts, reason and probability) is that the universe came into being as the result of an immensely powerful creative force outside time and space. But that's it. If you want to call that creative force God, fine. But we have no evidence whatsoever that this creative force is a personal, omnipotent, omniscient and benign God. Indeed, if we do postulate an omnipotent, omniscient and benign God, we have some serious problems in rationally explaining, for example, such phenomena as natural disasters and diseases that kill thousands of innocent people. And further testing questions about the efficiency with which He fulfils His presumed objectives.

3.1.2 Deism

It is easier to accommodate natural disasters and diseases under the deistic hypothesis. The deist believes that, having set the ball rolling, God then took his eye off the ball, so to speak. Natural disasters can perhaps be excused as relatively minor glitches in the running of the divine machine.

Nevertheless, the primary objection to deism is the same as for theism. We cannot, on the basis of the facts and reason, go beyond the factual statement that the universe came into being as the result of an immensely powerful creative force outside time and space. There is no factual evidence for a personal, omnipotent, omniscient and benign God.

3.1.3 Atheism

Given the starting point for this book, the obvious objection to the atheist's position is that the atheist not only has no explanation for the existence of the universe, he also argues that there is no explanation. It just is, and has to be taken for granted.

To refuse to accept as the cause for the universe a powerful creative force, outside space and time, requires faith. As we showed in exploring the question 'Why is there something rather than nothing?', the 'it just happened' proposition is not merely lazy thinking, it is not consistent with the facts (the universe came into being about 13.5 billion years ago), reason (we expect an effect to have a cause) and probability (given a choice between an explanation that is consistent with the facts and reason or a proposition that asserts there is no explanation, the former has to be more probable than the latter).

I can hear you saying that this critique of atheism goes too far.

> 'True, atheism can't explain some of the most fundamental questions about existence but that's no excuse for going off into the world or speculation or fantasy. After all, we know the material world in which we exist, the world of rocks and trees and hills and seas, the real world. Why don't we just accept it as given?'

Well, leaving aside the provocative question of 'given by whom?' (only kidding), there is a rather more fundamental objection to this extraordinarily ill-informed and intellectually indolent position. As the following pages will show, the world we see as real is a very partial and often false impression of reality; the real world, as revealed by science, is at least as improbable as belief in fairies, let alone God.

One last point. While there are many atheists who simply don't believe in God (a perfectly respectable point of view), there is a group of militant atheists who feel compelled to argue that belief in God is entirely irrational and unacceptable. They strut their stuff whenever and wherever they can, implying that anyone who does not agree with their materialistic, reductionist world view is ignorant or mentally challenged or both. It really is time to expose militant atheists for what they are, namely extremist, faith-based ideologues, with a fanatical determination to ignore the facts, abandon reason and opt for the least likely answer, or no answer at all, to life's most challenging questions.

We should put them in the same category as those theists and deists who assert the world was created literally in six days by a wholly benign, omnipotent deity – except, of course, that they are clearly less gifted with imagination.

(Objections: 10, 11)

3.1.4 Conclusion

So we conclude:

1. The universe, including time and space, came into being about 13.5 billion years ago
2. Whatever caused it to come into being was 'outside' time and space because the Big Bang included the creation of time and space
3. We know nothing for sure about the nature of the cause except what we can infer from the effect
4. What can we infer from the effect? Only that the cause was extraordinarily powerful (all the energy that exists within the universe was present at the moment of creation); orderly (the universe seems to be extraordinarily ordered at every level); and creative (the potential for all that exists was present at the moment of creation)

3.2 Why is there this particular something?

This may sound like another odd question so I'll try to explain.

We live in a very particular universe. First of all, it's a universe governed by rules. This in itself is an oddity we need to explore (see 3.4 below) but for now it is simply necessary to acknowledge that our universe is exceedingly well-ordered.

Here we need to discuss the nature of this ordering. What we find is quite extraordinary. We find that the universe seems to have been fine-tuned to ensure the development of stars and planets. If the rules were even slightly different, the universe, if it existed at all, would either constitute a chaotic sea of gas or would have collapsed in upon itself long before life emerged.

Martin Rees, former President of the Royal Society, described some of this fine tuning as follows:

> The ratio of the strength of electromagnetism to the strength of gravity for a pair of protons, is approximately 10^{36}. If it were significantly smaller, only a small and short-lived universe could exist.
>
> The strength of the force binding nucleons to nuclei is 0.007. If it were 0.006, only hydrogen could exist,

and complex chemistry would be impossible. If it were 0.008, no hydrogen would exist, as all the hydrogen would have been fused shortly after the big bang.

Just Six Numbers, Sir Martin Rees

It also becomes apparent that, if the rules were not almost exactly what they are, there would have been no possibility of the emergence of life or human consciousness.

> The expansion energy of the universe is closely matched to the attractive power of gravity. If gravity were stronger, the universe would have collapsed long before life could have evolved. If gravity were weaker, the stars would never have formed.

Ibid

Water is essential to the development of carbon-based life forms and yet the very existence of water with its odd characteristics is another example of an extraordinarily happy coincidence. Physicists have recently discovered that water has not just one quantum effect (which would remove many of water's special properties) but two, the second of which happily cancels out the other, thus restoring to water those peculiar properties on which life depends.

Fred Hoyle, the brilliant British astronomer, was puzzled by the abundance of carbon in the universe. To create carbon (on which all life as we know it depends), it was postulated that you needed three helium nuclei to collide. Given that the collision of two helium nuclei produced the unstable beryllium, there wasn't time for a third helium nucleus to join the party and create carbon before the unstable beryllium decayed. To explain the abundance of carbon Hoyle took the triple-alpha process which generates carbon from helium and calculated the precise resonance and spin that would be required for the process of carbon generation to work, i.e. that would allow enough time for the third helium nucleus to collide with the beryllium. His theory and calculations were proved correct. He had explained how carbon, the precursor of all carbon-based life forms, had come into existence. It was by a freak of good chance that the resonance of carbon was perfectly suited to the creation of carbon via the highly improbable beryllium route. It was so improbable that Hoyle wrote:

> Would you not say to yourself, 'Some super-calculating intellect must have designed the properties of the carbon atom, otherwise the chance of my finding such an atom through the blind forces of nature would be utterly minuscule. A common sense interpretation of the facts suggests that a super-intellect has monkeyed with physics, as well as with chemistry and biology, and that there are

no blind forces worth speaking about in nature. The numbers one calculates from the facts seem to me so overwhelming as to put this conclusion almost beyond question.'

> *The Universe: Past and Present Reflections*
> *Engineering and Science, November, 1981*

In *The Goldilocks Enigma*, Paul Davies set out to explore 'Why the universe is so uncannily fit for life'. At the cosmological level, he observes that, for life and then man to emerge, the universe had to have some very specific characteristics:

> The universe must be sufficiently old and cool to permit complex chemistry. It has to be orderly enough to allow the untrammelled formation of galaxies and stars. There have to be the right sorts of forces acting between particles of matter to make stable atoms, complex molecules, planets and stars.

> *The Goldilocks Enigma, Paul Davies, Penguin Books, 2007*

At the biological level, the requirements are no less particular:

> Among the many prerequisites for life – at least as we know it – is a good supply of various chemical elements needed to make biomass. Carbon is the

key life-giving element, but oxygen, hydrogen, nitrogen, sulphur and phosphorus are crucial too. Liquid water is another essential ingredient. Life also requires an energy source, and a stable environment…it took billions of years for life to evolve on Earth to reach the point of intelligence.

Ibid

Some examples of 'fine-tuning' have become less extraordinary in the light of more recent research but the list of fortuitous coincidences remains long. By some calculations, there are now 200 examples of how our universe seems to have been set up, first, to produce stars and planets; and then, to create an environment in which the highly improbable birth of life and the emergence of human consciousness could take place. If any of these factors was altered even by relatively small amounts, none of this, and none of us, would be here.

What are we to make of all this? We have the facts. As Hoyle put it, the universe seems to be a set-up.

But, before we jump to any conclusions, we must apply reason. What rational explanations are there for this evidence of fine-tuning?

First of all, it's worth pointing out that, if the universe wasn't set up as it is, we wouldn't be here to wonder about anything. In other words, if by some freak of chance, intelligent life evolved anywhere, it would be sure to find the universe was 'set up' to allow intelligent life to emerge. That thought may not be a complete answer to the mystery of all these fortuitous coincidences but it is a timely warning against jumping to conclusions.

Could all these instances of fine tuning really be no more than a long series of fortuitous coincidences? We can't dismiss this possibility out of hand but we can say the odds against them all being coincidences are astronomically high. Indeed, the odds against some of the individual examples being just good luck are on their own astronomically high. So, on our criteria, the 'it's all just coincidences' theory fails our probability test.

But before we decide we at last have evidence of design, if not a designer, we should consider a third possibility – the possibility that, for reasons as yet not understood, the universe itself has a sense of direction. This concept is not theistic. It does not require a God with a plan. It is derived entirely from observations of existence. Nor is it fully atheistic because it presupposes that existence itself has a purpose and is inexorably fulfilling itself according to drives inherent within it, drives which have, so far, created life and human consciousness. Also, importantly, this hypothesis is natural. Although it postulates a sense of direction and implies a purpose, this idea does not take us into the realms of the supernatural, any more than the emergence of life from inert matter, or the emergence of human consciousness from animal life are seen by us as supernatural.

We need to explore these ideas in a little more detail before we decide, according to our agreed criteria, which path to take.

(Objections: 12, 13)

3.2.1 The universe as a 'put up job'

As we have noted, there is a whole series of 'settings' in the laws of physics which seem to have been finely tuned to allow for the development of an element-rich stable universe, stable enough for life and human consciousness to develop.

At the same time, we must take into account that it took some 10 billion years for life to appear on Earth and 13.5 billion years for human consciousness to evolve so, if the universe was 'set up' to create life and human consciousness, clearly there was no sense of urgency. Anyone who takes the fortuitous 'settings' in the laws of physics as evidence of God and God's purpose (i.e. life and consciousness), needs to explain why it took so long and why, despite the favourable conditions, it proved so difficult for the universe to fulfil God's objectives. As far as we know, we are alone in the universe. Although there are trillions of planets, we have so far found no evidence of intelligent life anywhere other than on Earth. So, if the creation of life and man was

God's primary purpose in creating the universe, the deity, in fulfilling His purpose, seems to have been extraordinarily and inexplicably profligate in the allocation of both spatial and temporal resources.

Of course theists and deists might well argue that time and space are of no importance or relevance to God and that, in any case, His ways are inscrutable. Unfortunately, such a view requires a fairly substantial leap of faith and, as we agreed at the beginning, where we have a choice between a natural and a supernatural explanation of the facts, we will favour the natural.

We therefore conclude that, although the universe seems to have been set up for the emergence of life and consciousness, it has achieved its goals through a very long, discursive, evolutionary process, inconsistent with human concepts of efficient design.

3.2.2 It would have to be the way it is, wouldn't it?

It's certainly true that, if the universe wasn't set up as it is, we wouldn't be here to wonder about it. So, the argument goes, whatever the peculiarities of the 'set up', we shouldn't be surprised. On the surface this seems like a fairly convincing rebuttal of the 'universe is a put up job' proposition.

On rational consideration, it proves less satisfactory. Of

course we wouldn't be here if the laws of physics were not as they are. But, given that we are here, we should still be puzzled why the rules of physics seem to have been tweaked to our advantage. After all, we can easily imagine all the other possible universes in which the laws of physics had not been tweaked. We wouldn't be there but these others universes would exist, and the odds against our particular universe would be seen to be astronomically high.

This is a problem for the opponents of the 'put up job' proposition but, in response, they can deploy the 'many universes' theory. According to this theory, there are an infinite number of alternative universes, with laws of physics set to other, less favourable parameters. Our universe just happens to be the one with favourable parameters – and that's why we're here.

The problem with the 'many universes' theory is that there is no evidence for these other universes. Indeed, it is an integral part of the 'many universes' theory that their existence is inherently unprovable. Well, I have to say, if we are to reject God, a fairly simple concept to grasp, on the grounds that we can't prove His existence, we surely must be sceptical about a solution to a problem that requires us to presuppose an infinite number of unprovable universes, simply to explain this one.

3.2.3 The universe with a sense of direction

To date, there have been three extraordinary events in the life of the universe.

> First, there was the moment of creation, the Big Bang, when the universe, including time and space, came into being.
>
> Secondly, there was the birth of life on Earth, a small planet in a modest solar system on an outer arm of a spiral galaxy, the Milky Way.
>
> Thirdly, there was the emergence of human consciousness at the same address.

Without anthropomorphising, the universe seems bent on becoming conscious of itself. Yes, there are only three points on the scale so far but a trend seems to be emerging (energy/matter > life > human consciousness).

Given the popularity of the concept of evolution in biology and many other spheres, it shouldn't be too hard for anyone to apply the concept to the universe itself. There's no need to posit a God; but there is a need to observe and understand the universe and where it seems to be heading.

Of course some would argue, indeed it is the conventional

wisdom, that, given enough shakes of the dice, anything and everything can/will happen. According to this view, there is no sense of direction. All of evolution is explicable in terms of random mutations.

But it isn't true that, given enough shakes, anything can happen. In theory, if you have a truly infinite amount of time, it might be true. But there isn't an infinite amount of time. More to the point, there hasn't been an infinite amount of time. Here are a couple of experiments to ponder. I haven't done the maths but I think it unlikely that, even if you shook sufficient letters of the alphabet a million times every second for 13.5 billion years, you would, on any shake, get the complete works of Shakespeare. I also reckon that the odds against striking the white ball on a snooker table and downing all the other balls in one go are so great that, however hard you and all your descendants tried for however long, you would not succeed.

And there's another problem with the random dice-throwing proposition; not everything that could happen, happens. It is therefore perfectly reasonable to ask why, of all that could happen, a particular thing happened. And if there are several events that have happened which seem to have a common thread or a common driver, it is perfectly reasonable to explore the possibility that it is in the nature of existence to favour such events. And if those events which have a central quality in common are so momentous that, before they happened, they seemed unbelievably improbable, we are surely right, on the basis of fact, reason and probability, to consider whether together they might not indicate an existential trend, if not a purpose.

In other words, we should be sceptical of the 'it's just a matter of time and chance' argument and, if we discern a pattern or if extraordinary events occur, we should look for an explanation, not settle for 'that's just the way it is' or 'it's just what happens, given time'.

I would suggest that the three step-changes in existence, i.e. nothing to something, matter to life, life to human consciousness, are each even less likely as chance events than the Shakespeare and snooker ball experiments proposed above and that the facts strongly suggest that the universe is imbued with a sense of direction. It's not certain; but it is consistent with the facts; it is reasonable; and it is more probable than the alternative view.

(Objections: 14, 17)

3.2.4 The hindsight fallacy

There will be those who object to the general drift of the last three subsections. They would argue that, while it is true that there are many fortuitous factors that facilitated the creation of life, we should not be at all surprised, since, without those factors, we wouldn't be here to marvel at them. And they would suggest that the progression from inanimate matter to life and then to consciousness is simply a consequence of billions of years of shaking the dice, despite the problems with dice-shaking explored in 3.2.3.

Of course neither of these positions explains anything. Those who hold these views claim there is no explanation beyond 'Things are as they are because that's just the way it is'.

But I would argue they can get away with such lazy thinking and untrammelled complacency only because they are victims of the hindsight fallacy.

The hindsight fallacy consists of taking what has happened and then developing a theory of causation, underpinned by, if not wholly dependent on, the fact that things are as they are.

The best way to determine whether the hindsight fallacy is being employed is simple. Do not start from what is now and look back. Instead imagine you are in the past looking forward. If you were in the past looking forward, would you think it reasonable to say what happened next was inevitable or at least perfectly probable or, indeed, even possible.

Let's take an example. By a piece of paradoxical magic, you are back in the past before life emerged on planet Earth.

You exist in an entirely material, inanimate universe. You look around. You take into account the billions of years in which the material, inanimate universe has existed. Would you think it inevitable, or probable, or even possible that life would emerge? Of course, you have no idea what life is, so it's a tough question to answer. But, if you're honest, assuming someone told you what life was, you would probably think they were completely mad. The emergence of life was a qualitative leap in the nature of existence. Looking back, we can discover (more or less) how it happened. But, back then, looking forward, it was entirely unpredictable, even inconceivable.

Every day in the United Kingdom, we have an illustration of the mechanics of the hindsight fallacy. Each day, the weather people can explain exactly why the weather yesterday or a month ago was as it was. But they find it difficult to predict what it will be tomorrow and impossible to say what it will be in one month's time. Looking back, they can observe a persuasive cause and effect relationship; not so, looking forward. And this is just a question of predicting changes in a complex but homogeneous system, Life and human consciousness were step-changes in the nature of existence.

What is the point of all this? Simple. We take too much for granted. We know more or less how life started. But we don't know why, out of all the alternatives, life was what happened. And only a lazy, unscientific, uncurious mind, looking back from now, will say 'Well it happened; that's just the way it is.'

(Objections: 15, 16)

3.2.5 Conclusion

So we conclude:

> There is overwhelming evidence that the rules governing the universe were precisely calibrated from the beginning to allow the possibility of the creation of life and the development of human consciousness. This could mean that the universe was, as Hoyle said, a 'put up job'.

> For some, this invites belief in a God with a predilection for mathematics (the rules) and for interference (tweaking the rules) but lacking any sense of urgency (13.5 billion years to produce human consciousness) and a profligate attitude to real estate (an infinite universe). This view certainly requires an act of faith.

> An alternative, which takes account of the facts but avoids any leap of faith, is the theory that the universe itself has a sense of direction. It has moved from inanimate matter to life; and from life to human consciousness. At each stage, it has not abandoned the raw material of existence but has refined it and added to it. The inanimate became

animate; the animate became conscious.

Because we favour the natural explanation over the supernatural explanation, at this stage, we favour the 'sense of direction' hypothesis.

Note that both of these hypotheses (God or the sense of direction) conflict with the commonly held secular view that things are as they are because that's just the way it is. But such complacency is possible only by employing the hindsight fallacy.

There will be many readers who do not feel comfortable with these conclusions. One reader will feel there is sufficient evidence for belief in the God. Another that these conclusions invite types of speculation (e.g. theism or deism) in which they refuse to engage. To both types of objector, I would say that we are looking for answers based on fact, reason and probability and I would contend that neither of you can present a set of conclusions that meets these criteria as well as the alternative (sense of direction hypothesis) outlined above.

To those who take the conclusions as an invitation for a leap of faith in God, we have to say there is no foundation, in fact or reason, for such a leap.

To those who wish to cling to the random, dice-shaking, materialistic explanation of existence, we have to say much the same, except that, on the balance of probability, the atheist is on even weaker ground than the theist, if only because the atheist is unable to offer any answer at all to the first and most fundamental of questions: **'Why is there something rather than nothing?'** And then answers most of the subsequent questions by saying: 'That's just the way it is' – which is rather like the 'Because I say so' answer given by a lazy parent to the curious child who asks 'Why?'

3.3 What exists?

We now need to explore the nature of what exists. After all, if we are to draw any conclusions about the origins and, if there is one, the purpose of existence, clearly we need to know as much as possible about the nature of existence.

First, in day to day living, we must acknowledge we are trapped within our senses (seeing, hearing, smelling, tasting, touching). We have a strictly sense-limited view of what exists.

Of course we know, mainly through experiments with scientific instruments, that there is more to existence than we can experience through our senses.

The human eye is limited to perceiving light in a fairly narrow range of wavelengths (390 to 750 nanometres). Within this range, we see the following colours: violet, blue, green, yellow, orange, and red. But the electromagnetic spectrum is far wider than that. We cannot see ultraviolet radiation (which has shorter wavelengths) or infrared radiation (which has longer wavelengths), although we can experience some of these wavelengths through another of our senses, touch (ultraviolet radiation when the sun burns our skin; and infrared radiation when we feel the heat of a fire).

Then again, there are sounds beyond the human

auditory range. A young adult can hear sounds in the range 20 to 20,000Hz. Sounds below 20Hz (infrasound) and above 20,000Hz (ultrasound) exist but are outside our hearing range.

So, through our senses, we are aware of just some of what is 'out there'. If we want to know the true nature of material reality (the stuff that we think of as 'real', as opposed to figments of our imagination which are insubstantial and 'unreal'), we need to turn to science.

The quest for the basic component of matter has a long history. We can begin with Democritus (460–370 BC) who formulated the theory of atoms, tiny units of matter that were indivisible and made up everything. This theory allowed us to continue to believe we knew what matter was, even if its individual atomic components were too small to see. It was still 'solid stuff'.

It was not until the late 19th and early 20th centuries that matter ceased to be quite as solid as we thought. The discovery of the electron and the atomic nucleus led to the view that matter is composed of electrons (which are negatively charged), protons (which are positively charged) and neutrons (which are neutral). Today, we now know that protons and neutrons can be broken down into quarks; and electrons are now considered one of a particle family called leptons.

Perhaps the most interesting feature of the atom, as we currently understand it, is that it is composed almost entirely of empty space. For example, the most common atom in the universe is the hydrogen atom. The hydrogen atom consists almost entirely of space. All the particles in a hydrogen atom constitute only 0.0000000000004% of the atom. The rest is emptiness.

It is difficult for us to grasp but, when Hamlet pleads:

> 'O that this too, too solid flesh would melt
> Thaw and resolve itself into a dew!'

Shakespeare, Hamlet, Act 1, Scene 2

science would grant his wish and apply it, not just to flesh, but to tables, chairs, rocks, planets and stars. From the point of view of particle physics, there is almost nothing there, at least very little matter.

At the same time, science also tells us that there are extraordinarily powerful but intellectually understandable forces at work which give this miniscule amount of matter sufficient coherence to create and sustain stars, planets, rocks, chairs, tables and our too, too solid flesh.

So we have to accept that we live in a largely immaterial universe of powerful and invisible forces, governed by intellectually comprehensible rules.

That's the truth. And that is science, not religion, speaking.

(Objection: 18)

3.4 What are rules?

We tend to take it for granted that we live in an ordered universe and yet it is extraordinary that every particle of existence anywhere in the universe behaves in accordance with mathematically definable rules. So it's well worth asking: 'What are rules?'

It's not an easy question to answer but we can define the question more precisely. In this context, we're thinking about the rules that govern the material world – rules like gravity.

Newton's law of gravity states that any two bodies in the universe attract each other with a force that is directly proportional to the product of their masses and inversely proportional to the square of the distance between them. Why should all matter behave according to mathematical formulae?

It's quite tempting to think of rules as abstract, immaterial commands which somehow impose their will on material things – and that is a possibility. But it's also possible to think of rules as inherent in material objects.

Either way of thinking is problematical. If material objects are subject to immaterial, abstract commands, where are these commands and how do they control things? If the rules are somehow embodied in things, is it credible that every atom somehow knows exactly how to behave in relation to every other atom in the universe? Where exactly does that atom carry such information?

Gravity is a particularly interesting example because, although we know how it operates and, post-Einstein, have an even more refined formula to predict its effects, we still are not certain what it is. Some scientists believe there is a graviton particle which is the mechanism whereby gravity operates but, even if this is so, it won't explain why it acts consistently according to a mathematical formula.

Of course, some would say that both ways of understanding rules, as set out above, are misguided. The only rules that exist are those devised by the human mind in order to understand and, in some cases, control nature. There are no rules 'out there'. They are simply mental constructs derived from our observation of nature and then imposed by the human mind on nature. This is a powerful argument. After all, by definition, all the rules we know are constructs of the human mind.

On the other hand, the human mind cannot create rules arbitrarily. The rules have to work. Nature has to conform to them. In other words, the rules *are* 'out there'. We can only discover rules if there are patterns in nature to which we can successfully apply the rules.

In deciding the fundamental questions about the nature of existence and our place in it, we need to take account of this feature of the universe. The whole of science attests to the 'orderliness' and 'comprehensibility' of the material world. As with many other aspects of existence, we take this feature of the universe for granted but, given that there are always far more possibilities for chaos than order, we have another anomaly to explain or, put more positively, another piece of evidence to guide our conclusions.

4. HUMAN EXPERIENCE

In any attempt to answer fundamental, existential questions, the best source of information to mine, to explore and on which to base our conclusions must be human experience. We need to understand the human predicament; to set ourselves in the context of what we know of the universe; and, as best we can, to evaluate the meanings that humanity has attached to existence.

This section will also help us to understand the concept of God. It doesn't matter whether we are of the opinion that man created God or vice versa, probably the best indications of the nature of the concept of God will be found by studying man. After all, either man is God's creator or, as far as we know, God's finest creation.

Alexander Pope said:

> 'Know then thyself; presume not God to scan.
> The proper study of mankind is man.'

Pope's Essay on Man

With apologies to Pope, I'd make a rather different point.

> To know your God, best that you scan mankind
> in hope to glimpse the nature of God's mind.

In this section I will briefly review major areas of human experience to see whether they can provide any help in answering the questions posed in the previous sections.

The step change from animal to human consciousness brought with it great blessings and a great curse.

On the positive side, it allowed us to think conceptually, to reason, to control and exploit our environment, to be creative and to imagine. It allowed us to transmute animal drives into more subtle, refined, and complex types of experience, mediated by our creativity and imagination.

It has also given us morality, a sense of right and wrong, a conscience.

On the negative side, it compels us to confront death, not just as an event at the end of life but as an inescapable truth throughout life.

In the following sub-sections we will explore the human predicament and try to determine what it can tell us about human nature and the nature of existence.

4.1 Half angel, half beast

Placed on this isthmus of a middle state,
A Being darkly wise, and rudely great:
He hangs between; in doubt to act, or rest;
In doubt to deem himself a god, or beast;

Created half to rise, and half to fall;
Great lord of all things, yet a prey to all;

Alas what wonder! Man's superior part
Uncheck'd may rise, and climb from art to art;
But when his own great work is but begun,
What Reason weaves, by Passion is undone.

Pope's Essay on Man

It's not easy being human. We have imagination. This means that we have hopes; we have aspirations; and we have fears. We ride on the back of hope as we canter inexorably and helplessly towards death.

We are a body and a mind. The body is our means of contact with the world. Our mind allows us to understand the world. Our body, the source of all our physical pleasure and pain, is indisputably animal, with all the gross material processes of eating, digestion, excretion, copulation and procreation. Our mind which functions primarily in an immaterial world of ideas, impressions, emotions and imagination, continually asks challenging questions, such as 'how?' (which is often relatively easy to answer) and 'why?' (which is generally trickier).

Although individuals, each with our own sense of self, we have the need to form relationships and societies. We have a conscience. We learn or are taught to treat others as we would have them treat us, to empathise. But our capacity for empathy conflicts with our inherent self-centredness, creating intractable moral conundrums.

Let's face it – we're in an appalling and ridiculous predicament. It's not surprising we need to spend about a third of our lives asleep in a forlorn attempt to sort it all out or to get away from it all.

We're a bit like the small dinosaur that is only halfway through the process of becoming a bird. If we believe in evolution, man is almost certainly unfinished business. As the small dinosaur must have thought, as she struggled to survive, with long, impractical, jointed appendages and nascent feathers but still unable to fly, 'This cannot be what I'm supposed to be. I must be on my way to something else – and hopefully something better'.

4.2 The Arts

One way we attempt to deal with this 'half angel, half beast' conflict is through the arts. We take the world as it seems to be and try to infuse it with significance, with value, with meaning.

The act of artistic creation emulates what believers think of as the act of divine creation. We take the raw material of our experience and we select, we organise and we create. Every act of creation is a miracle, in the sense that the resultant whole is greater than the sum of its parts; somehow the whole possesses emergent properties which cannot be attributed to any one or any incomplete combination of the constituent parts.

Once again, we tend to take this feature of experience for granted. We need to ask how it is that the whole is greater than the sum of its parts.

Let's take the creation of a poem. We select words; we organise them; we create.

Agreed?

Yes, but I've omitted the very first act. We have to have a thought, an idea, something we want to express. The thought is pure and complete or, at least, it should be; the expression of the thought, however hard we try, is almost always imperfect. Occasionally, we may find the perfect expression of our thought. We find the precise words, with the entirely apt connotations, with the exact sonic and rhythmic qualities that, when suitably organised, perfectly express our thought. When it happens, it's an odd mixture of purpose, technique and luck, infused by the mystery of the origin of the thought at the heart of the poem. All the words in the poem have been used countless times before but this particular arrangement of them can be sublime.

A painting stops the world for a moment. Unlike a poem which exists in time (with a beginning, a middle and an end), a painting is caught in an eternal now or, perhaps more accurately, exists outside time. What matters is how all the parts of the picture fit together and relate to each other. This is no place for the temporal sequence of cause and effect. All of it simply is. We are invited to observe the arrangement and significance of subjects, shapes and colours. We can, if we so wish, infer the artist's intention and absorb and assess our intellectual and emotional response to his creation but the real value lies in the immediacy and, at the same time, the timelessness, of our viewing experience.

Music, with or without words, is one of the most powerful forms of art. For many people, music engages more directly and can move more deeply than any other art form. Of all the arts, music transports people's consciousness away from self-consciousness more directly and effectively than any other.

Creative works give us the opportunity to experience the products of minds other than our own. Artists strive to express what they have to say as best they can. In one sense the audience is passive and the artist is the active one, the creator; but, in another sense, the artist is entirely absent (his/her work is done) and we, God-like, outside the artist's time and space, can appreciate and comprehend the artist's creation as a whole. Drama is probably the most powerful example of this process. We become intensely and intimately involved in the work and yet, crucially, we are outside it, empathic observers. This phenomenon leads us to ponder a peculiar aspect of human consciousness.

All art forms seem to have one other thing in common. They provide the audience with an opportunity for self-less consciousness, i.e. consciousness without consciousness of self. In fact they enhance our consciousness while suppressing our self-consciousness. The paradoxical common expressions for this experience are 'It takes you out of yourself' or 'We become lost in the work'.

The two features of existence noted above, emergent properties and self-less consciousness, merit further consideration.

4.2.1 Emergent properties

We often observe in instances of artistic creation and, indeed, in many other fields that the whole is greater than the sum of its parts.

Once again, we tend to take this feature of experience for granted. We need to ask how it is that the whole is greater than the sum of its parts. The poet puts the breath of meaning into the inert, denotative words he chooses. The philosopher defines new, subtle and refined concepts with words which on their own have no right to aspire to such intellectual heights. The novelist creates characters and worlds as a builder creates a palace from a pile of bricks. All these feats seem miraculous.

No, you say. It's not a miracle. And you are right. It's not supernatural; it's not magic. But it is extraordinary. Whatever it is that makes the whole greater than the sum of its parts is immaterial. It is a thought, a sentiment or a design. It seems to be an inherent feature of our universe that more will come from less. Evolution itself, in physics, in biology and in human consciousness, provides abundant evidence of the capacity of existence to manifest emergent properties, properties that, with the help of concepts and organization, transcend the components of the whole and can be explained only as an integral creative function of reality.

(Objection: 19)

4.2.2 Self-less consciousness

It's important to emphasise that, in using the concept of self-less consciousness, we are simply describing a human experience.

For much of our lives our understanding of our experience is self-centred. It has to be. I am confined within a body which is my interface with the rest of existence. We have had to accept that our direct experience of existence is self-centred (i.e. mediated through our senses, or sense-limited, as I called it). If I am cut, my body bleeds and I feel pain. In planning my life, my relationship, my business and my leisure, my self is at the centre of it all.

But there is a form of consciousness which is rich, perhaps richer than other experiences, in which self plays little or no part. In some religious experiences, in appreciation of art and, oddly, in some manifestations of love, the self fades away, enabling the emergence of what we might call self-less consciousness.

Some make a religion out of this phenomenon. Buddhism, in particular, tells us that the self is a delusion, the source of all human suffering, and should be abandoned.

While holding Buddhism in the highest regard, we are not suggesting that course of action. We are not, for example, prepared to take a leap of faith and subscribe to the notion of rebirth and reincarnation, given the absence of evidence to support the hypothesis.

But equally, on the grounds of fact and reason, we will give due weight to the phenomenon of self-less consciousness and will take it into account when drawing up our list of final conclusions.

(Objection: 20)

4.3 Love

Love is a large subject but one which we cannot ignore if we are pondering the human experience in the hope of reaching conclusions about what exists and its significance.

There have been many attempts to define love. Some have tried to list the different forms.

The ancient Greeks had many words for love but four words give a good indication of what they saw as the scope of love:

- Agape (ἀγάπη): selfless love, love that demands nothing in return
- Eros (ἔρως): physical, sexual love
- Philia (φιλία): non-erotic love between equals, the love between friends, love for one's family, love for one's community
- Storge (στοργή): the love of parents for their children

In English we have one word for love but distinguish the different types of love with adjectives: erotic love, romantic love, platonic love, familial love, humanitarian love, spiritual love.

Others have set out to identify the key components of love (e.g. intimacy, commitment, passion) and have then tried to show how all the different types of love are made up from these key components in varying proportions.

However we define it, there is no doubt that love is the most powerful positive emotion human beings experience. It is what holds together marriages, partnerships, families, communities, societies and humanity.

It has one particular distinguishing feature. It requires the one who loves to give the highest priority to something other than self. Even in erotic love, where there is an undoubted element of self-gratification, the primary focus is on the other, rather than the self.

It seems that the greatest of human joys comes from the subordination or denial of self in achieving union with something other and greater. Christians encapsulate this notion in the paradox of describing service to God as 'perfect freedom'.

(Objection: 21)

4.4 Morality

Let's agree that man has a moral sense. He may not behave well but a sense of right and wrong seems deeply embedded in the human mind. We don't all agree on what is right and what is wrong but almost everyone has a view on good and evil and most subscribe to some kind of a moral code. Even the thief has honour amongst his peers; even the murderer was good to his mother (except of course in cases of matricide).

So where does this moral sense come from?

David Hume asked how we move seamlessly from 'is' to 'ought' in discussions of human affairs, without having a clear rational basis for the transition. In other words, what is the basis for human morality? It's another tough question but, tough or not, if we want to understand the nature of human existence, we have to address it.

In simple terms, humanity has come up with four suggestions:

God

God is a perfect validator for a moral code. He is generally conceived to be both omnipotent and good. He therefore has the power, and one might say the right, to decree a moral code. The only snag is that it all depends on whether or not you believe in God.

human authority

A government; a lawgiver; a teacher, a religious organisation, an ideologue or a philosopher can propose and, in some cases, impose a moral code. The essential key to human authority as a basis for a moral code is the willingness of the individual to accept the authority of the authority.

reason

Kant was one of many who have argued that reason leads us to discern and follow a moral code. The argument for the use of reason is based on empathy which assumes that all humans are of equal value and that therefore, as Jesus said, 'Do unto others as we would have them do unto you', or as formulated by Kant in his categorical imperative: 'Act only according to that maxim whereby you can, at the same time, will that it should become a universal law'. The problem here is that the empathy that justifies the moral code is in itself a moral principle. In other words, accepting that others have the same rights as ourselves is a moral principle which is not based on reason.

experience

> It is arguable, though far from certain, that fully understanding our experience of life inevitably leads us to a moral code, through a process of either enlightenment or enlightened self-interest.

It's a fascinating subject but one which we cannot allow to detain us for too long.

Apart from the divine authority of God (predicated on faith rather than reason), none of the alternative sources of morality is entirely convincing. They all depend, to some extent, on the argument for equality in the value of human lives. They therefore falter immediately if anyone denies that equality. Some might say that it is self-evident and rational to assert such equality – but it is neither. If we look at the facts and apply reason, we will find that history and the structures of all contemporary societies demonstrate that all human beings are not equal. Only the faith-based morality of God survives the test by asserting we are all equal in the eyes of God. By any other yardstick (charisma, competence, diligence, dynamism, generosity, humility, intelligence, integrity, kindness, physical strength, wealth, wit, etc.), it's pretty obvious that we aren't all equal. Society can insist that all people should be treated as equal; we might agree with it; ***but we might not*** and, such disagreement would not be irrational. So reason (*pace* Kant) is a pretty shaky foundation for morality.

That said, we should perhaps focus on a more interesting question: given the selfish drives in man, why do we have a moral sense? Why do we have a sense of right and wrong? Why do we have a capacity for empathy? Why do we feel guilt when we transgress? Why do we have a sense of release and empowerment when we do the right thing?

The answer could be that, having sourced and absorbed a moral code from human authority, reason or experience, we have negative feelings when we fail on our agreed moral standard and are proud and uplifted when we meet it. Such feelings would be reinforced by social disapproval of failure or approbation of success.

Against this view is the fact that we can feel guilty when no one knows we have transgressed, even when we can be sure no one will ever know. It seems we can be ashamed of being less than we think we should be. We have a moral sense; we aspire to behave according to our moral code; and we are ashamed if we fail.

We should also note something else. All moral codes require us to deny ourselves to some extent. Selflessness is a key part of moral consciousness. So we could say that, if we aspire to be better than we are and a key part of being better is selflessness, then we (individuals profoundly conscious of our self because of our separation from the rest of existence by our bodies) paradoxically aspire to some degree of selflessness.

That would suggest that the consummation of morality is love, the love which requires us to subordinate the self to the other, whether that other is God or gods, another individual or humanity as a whole. At the same time, this subordination of the self to the other promises us enhanced (but largely self-less) consciousness.

This analysis is consistent with all the world's great religions – all of which promote, in one way or another, a degree of self-denial. Buddhism ascribes all suffering to the illusion of self. Christianity reveals a God 'whose service is perfect freedom'. The word Islam means submission to the will of God.

(Objection: 22 and 23)

4.5 Conclusion

The human condition (an animal with self-consciousness, a moral sense and awareness of its own mortality) is essentially appalling and ridiculous.

Locked in a material, animal body but with the mental capacity to understand existence and enter worlds of imagination, we find ourselves in a conflicted situation, probably best explained by assuming we are 'work in progress'.

Our chosen way to explore and ameliorate the human condition is through art.

A key feature of artistic endeavour (and of many other areas of experience) is the phenomenon of emergent properties. The whole becomes greater than the sum of its parts. We need to consider the significance of the propensity of reality to enable elements, when suitably arranged, to exceed or transcend their apparent combined potential.

We also note that all art forms seem to have one thing in common. They provide an opportunity for self-less consciousness, i.e. consciousness without consciousness of self. This is a form of consciousness which is rich, perhaps richer than other experiences, and one in which self plays little or no part.

The most powerful positive human emotion is love. Love requires the one who loves to give the highest priority to something other than self. Given the human predicament, essentially a limitless conscious mind trapped in a temporal and temporary animal body, it is both paradoxical and informative that the greatest of human joys involves the subordination or denial of self.

The phenomenon of self-less consciousness (as experienced though forms of art, social interaction, friendship and love) appears to constitute a form of being, less constrained by time and space. Such experiences are immaterial forms of awareness; they are not subject in themselves to any law of organic decay; they are not, in any meaningful sense, located on spatial coordinates; and, while they are certainly temporal in one sense and material in origin, they often seem to break out from the normal passage of time to touch something timeless.

Man has a moral sense. Selflessness is a key part of moral consciousness. It is paradoxical that individual selves, clearly separated from the rest of existence by their bodies, should aspire to selflessness. Given that both morality and love require some degree of selflessness, we might even speculate that both the origin and the consummation of morality is love, the love which requires us to subordinate the self to the other, whether that other is God or gods, another individual or humanity as a whole.

While admitting the suggested connection between selflessness, love and morality is hypothetical, none of this should be seen as drifting off into mysticism or the supernatural. These conclusions are simply an attempt to identify the essential nature of the group of experiences which are generally felt to be the most important and positive in human lives.

5. GOD?

We need to consider the concept of God in some detail simply because, to date, He seems to be humanity's most popular answer to the questions we have been pondering.

5.1 Should we take the concept of God seriously?

Another difficult question!

On the negative side:

a. there is no scientific proof of the existence of God, any more than there is scientific proof of the existence of dragons or fairies. If you are going to have faith in entities for whose existence there is no scientific evidence, there's no limit to what you can choose to believe in

b. if God exists, why does He hide himself?

c. why does He demand faith? What's so special about faith? It is, after all, possible only if you abjure facts and abandon reason, not practices of which we approve in any other aspect of life

d. the conception of God as wholly benign, all-powerful and all-knowing is riddled with contradictions (see subsections 5.2 and 5.3)

e. it is unclear why an all-powerful God would want to create an inferior species whose primary duty should be to worship its creator

On the positive side:

a. throughout human history, belief in gods or God has been widespread

b. from the Egyptian Pharaoh Akhenaten on, monotheism has achieved remarkable popularity

c. there has been a series of prophets who have claimed to provide revelations of the nature of God

d. despite scientific advances, probably most of the world's human population believes in God and many claim to have had convincing personal experiences of the presence of God

e. if true, God provides a simple (some would argue, the only) explanation of why the universe exists, why we are here and why we are as we are

Using our criteria for answering key questions, we

have to conclude that while there is powerful, factual evidence for the existence of a powerful creative force, outside time and space, which initiated an intellectually comprehensible universe and which probably imbued it with a sense of direction, there is insufficient evidence, on the basis of fact and reason, for the existence of God as presented in the great monotheistic religions.

At the same time, people's propensity for faith in God or gods and the similarity in the key messages conveyed by these religions suggests at the very least that humanity aspires towards something greater and better than itself. It remains unclear whether this drive is simply a human aspiration or whether it is recognition of and communication with an external, objective reality.

While we do not find in any religion sufficient justification for faith in God, it would be unreasonable to dismiss out of hand the revelations of the Buddha, Jesus Christ, Muhammed and other great religious leaders. It seems at least possible that each of these men glimpsed, with varying degrees of accuracy and perspicacity, the nature of an external, powerful, creative and moral entity; or were, at the very least, inspired by a vision of a higher level of existence.

Furthermore, while we may not have faith, we concede that the phenomena of Buddhism, Christianity and Islam, the religious tendency with its insistence on moral precepts, is not entirely satisfactorily explained in terms of wishful thinking or fear of death.

(Objection: 24)

5.2 What could we know of God?

If, for a moment, we assume God exists, what could we know of Him?

There would seem to be two main sources of information:

a) the nature of creation

By His works shall ye know Him. If the universe is God's handiwork, then, although He has chosen to hide from us, we should be able to deduce something about the creator from what He has created.

b) revelation by prophets

There has been a long tradition of wise men and prophets who have given us the benefit of their understanding of God. Unfortunately, there has been some variation in their accounts which range from the vengeful, essentially malign God of the Old Testament to the compassionate, all-forgiving God of the New Testament.

In any case, while we respect and should take account of the revelations of the prophets, they cannot be a determining factor in our quest for truth, since they demand faith and do not meet our criteria of fact, reason and probability.

That leaves us with the possibility of inferring the nature of God from our experience of His creation.

5.3 Questions about the nature of God

5.3.1 Creative and powerful?

If God exists, there are two qualities we can ascribe to Him with confidence. He is creative. He is powerful. Creation attests to both. For those who believe, God created the universe 13.5 billion years ago. From an infinitesimal point (a singularity), all that exists emerged. We can't say what existed before the Big Bang because time came into being with the Big Bang. There was no 'before'. All we can say is that the cause of the Big Bang must have existed somehow outside the Big Bang, and therefore outside time and space. (Outside, being a spatial concept, is not the right word, but it will have to do.)

Of course, there are some very eminent scientists who, in their determination to present an atheistic explanation of reality, are desperately trying to prove that the universe was created out of nothing. Put very crudely, they seem to be arguing that, because -1 and $+1 = 0$, then 0 can generate -1 and $+1$. We wish them success but we would point out that, even if they succeed, they would still need to explain why it was in the nature of the primal 0 to express itself as -1 and $+1$.

In other words, in a profound sense, they really haven't understood the question.

(Objection: 25)

5.3.2 Benign?

All the world's monotheistic religions see God as the source of all good.

There is, however, a problem. If God created everything and if there is evil as well as good in the universe, He must be responsible for both good and evil. Most religions deal with this problem, either by opting for dualism (accepting there are good forces and bad forces in the universe) or by positing a fallen angel who has rebelled against God.

The problem of the wholly good God of monotheistic religions is compounded by insistence on God's omnipotence and omniscience (see sections 5.3.3 and 5.3.4).

There is also a question of what we mean by 'good' when we say God is good. God's morality seems to have been dependent on the custom and practice of the times in which religions developed. The God of the Old Testament is basically a tribal God. He commands genocide to provide his chosen people with a land of their own. He wipes out men, women and children in Sodom and Gomorrah because of the depravity of some of their citizens. The Prophet Muhammad taught of a God who was the Compassionate, the Merciful, but the Prophet was still able to accommodate in his revelation of the nature of his benign God bloody battles to establish his faith and the enslaving or extermination of Jews and others who failed to support him.

The God whom Jesus came to reveal represented what most moral philosophers would agree was a morality of a much higher order. This God was a God of love and forgiveness, a God of infinite compassion, who offered redemption to all. By the time St Paul had finished his work, Christianity was a truly inclusive global religion. The tribal God of the Jews had become the Christian God of humanity.

Assuming the existence of God, man's understanding of his creator has evolved substantially over the centuries. Viewed from the moral standards of the 21st century, the God of the Old Testament must be seen as morally inferior to an averagely 'good' human being. His obsession with obedience, His demands for love, His enthusiasm for genocide make Him entirely unacceptable as an object of worship. He is a self-obsessed, manipulating sociopath. If a believer believes God is good, it is irrational to believe in and worship such a God.

The God of the New Testament is different in kind. He accepts all humanity and forgives all its sins. This is a God more in tune with modern morality; indeed quite possibly considerably in advance of it.

The God of Islam, as revealed by Muhammad, seems to fall somewhere between the God of the Old Testament and the God of Jesus. The God of Islam seems to be, by origin, a tribal god of the medieval desert but with aspirations to embrace all mankind in a greatly enlarged tribe.

As a mental exercise, it is interesting to compare the morality prescribed by the great religions with the morality embedded in human rights legislation in the 21st century. Unhampered by revelation or dogma, it could be argued that human rights legislation has, in many ways, set a higher moral bar than most organised religions.

Fact, reason and probability would lead us to conclude that, if there is a God, He is powerful (but not omnipotent), clever and wise (but not omniscient) and generally, though not invariably, benign. See below, and also 5.3.3 and 5.3.4.

5.3.3 Omnipotent?

A common feature of all monotheistic faiths is that God in omnipotent. If He is omnipotent, He has to take responsibility for natural disasters and diseases which can kill hundreds or thousands of innocent people.

The defenders of God respond by telling us the ways of God are inscrutable and beyond our understanding. But that's not good enough, not good enough if we are to hold to our criteria of facts, reason and probability.

It is not reasonable to argue that a wholly benign and omnipotent God would permit the slaughter of innocents by natural forces and phenomena over which He presumably has absolute control. Applying reason, we have to conclude that either God is not wholly benign or God is not omnipotent, or He is neither.

Given we must abandon at least one of his attributes, again applying reason and probability, we should sacrifice God's omnipotence, rather than His benevolence. After all, our world, like our lives, is a wonderful gift. It may not always turn out to be wonderful for each and every one of us but, on the whole, if there is a God, most of us would agree He seems, on balance, good rather than bad. And if He is good, He can't be omnipotent.

5.3.4 Omniscient?

There's also a problem with God's omniscience, at least for the three great monotheistic religions that came out of the Middle East.

All three religions teach that all the evil in the world ensued from Eve and Adam's eating of the apple of the Tree of Knowledge of Good and Evil.

The first problem is simple enough. The God of the Old Testament makes it clear that **good** is essentially obedience to the will of God. By telling Adam and Eve not to eat the apple, He had already given them knowledge of good and evil. Eating the apple was bad because He forbade it. His reaction when Eve and Adam ate of the apple therefore seems excessive, misjudged and unfair. He was not upset that they had acquired a sense of morality; He was outraged simply because they had disobeyed.

The second problem concerns free will. God gave man free will. That means we have a choice of doing good or doing evil. So far, fair enough. But, if God is omniscient, He knows what we are going to do. Is it consistent with a benign and omnipotent God that He would create humanity, knowing that, when given a choice, we would disobey His command and, as a result, would be condemned, through the generations, to have to wrestle with temptation and face mortality? Why would He want to do that? And, in Christianity, by what twisted logic would it make sense to send His son to be crucified by fallen man in order to redeem fallen man. Surely crucifying His son was even worse than eating the apple? In any case, is it not the offender, not the offended, who redeems himself through suffering punishment. Or was the crucifixion God's apology for putting us in the ridiculous position of the human predicament (see 4.1)?

If we want to maintain a belief in a creative, powerful benign God, applying reason and probability, we have to forgo His omnipotence and omniscience. If God forgoes omnipotence and omniscience, He no longer needs to explain why there are natural disasters and diseases in the world, a great relief to those who find the existence of such evils a serious impediment to faith in God. On the other hand, a God stripped of omnipotence and omniscience will probably prove unsatisfactory to those of faith in the great monotheistic religions.

5.3.5 The Evolution of God

The concept of God is endlessly fascinating to the rational mind.

It involves faith in an invisible being of immense creative power who takes a profound personal interest in each and every human being.

Most atheists would argue this is the product of wishful thinking and fear of death. We find ourselves alone in the world, connected to but at the same time separated from the rest of existence through our individual bodies. We have a mind which is capable of the most extraordinary accomplishments but dependent on the brain, an organ which we know will die with the rest of our corporeal components. It's scarcely surprising that we find the prospect of death (our own and that of loved ones) deeply disturbing. To deal with this existential problem, so the atheist argues, mankind has invented a fantasy of a supreme being and an afterlife, simply because the prospect of an impersonal accidental universe and our permanent obliteration at death is too hard to bear.

Against this, to be fair, we must set the religious experience of millions of human beings over millennia. If religion is fantasy, it is a fantasy with the capacity to inspire great acts and great art. We should also take account of the prophets, many of whom have claimed to have had personal experience of God, or his angels.

If we were to leave it at that, the factual, rational mind would probably side with the atheist, albeit reluctantly, or at least opt for agnosticism. But we need to delve deeper.

There are two features of the concept of God which merit further thought. First, there are similarities in all the world's great religions. They all talk of a God who is essentially benign. Words such as compassionate and merciful are favourite epithets. Most tell us that, to achieve the good life or enlightenment, we must be selfless, deny our selves, subordinate our selves to a higher purpose – in other words, transcend the corporeal limits (appetites and demands) of our bodies.

Secondly, despite the similarities, the concept of God has changed over the centuries – it has, if you like, evolved. Of course, the deists and theist would argue that God is immutable; it is simply our understanding of God which has evolved and, hopefully, improved. For the atheist or agnostic, it seems more likely that the concept of God has been refined by man as we have learned more about the nature of existence and thought more deeply about our place in it.

We must also take into consideration the drive in man to find meaning in life. Religion is an obvious example. Whether a religion is true or not, it provides its adherents with a sense that their lives have a point, a significance, a value. What is interesting here is not the outcome, but the drive itself. **Man is a meaning-seeking creature.** Sometimes he finds meaning; sometimes he makes meaning; and sometimes the boundary between the two (making and finding) becomes blurred. But, however meaning arises, the acquisition of meaning seems to be a fundamental and inherent requirement of human consciousness.

If we apply our factual, rational, probabilistic criteria to the above account, we conclude that there is insufficient evidence to support the existence of a fully formed, benign, omnipotent, omniscient creator but that the similarities in the various concepts of God across different cultures, the evolution of man's conception of God, and man's innate need for meaning cannot be wholly dismissed as the product of fear or wishful thinking.

The atheist would disagree, explaining all conceptions of God and man's search for meaning simply as symptoms of existential angst but this position is not entirely rational. The problem is not that the religionists necessarily have the right answer; the problem is that the atheist seems to dismiss as insignificant too much of the human experience and the known nature of the universe.

The religious experience of millions, the proven effects of that experience on the human spirit and the drive in man to give life meaning, do not, in our view make the existence of God probable but it is probable that **there is some form of reality, some mode of existence, to which all the above relate, regardless of whether or not that form and mode are created by man or exist independently of man.**

We have considered the atheist's view that fear of death is the primary driving force behind religion. While there may be some truth in this explanation, it is weakened by the inevitable fear of the hereafter which, for the devout sinner, is far more frightening than mere death.

> 'To die, to sleep -
> To sleep, perchance to dream. Aye there's the rub
> For in that sleep of death what dreams may come
> When we have shuffled off this mortal coil
> Must give us pause.'
>
> *Shakespeare, Hamlet, Act 3, Scene 1*

Nevertheless, fear remains the cornerstone of the atheist's deconstruction of the 'religious tendency'.

There is another possibility that is worth throwing into the pot. The apparent predisposition for mankind to adhere to religions might indicate glimpses of the future, an awareness of where we, as a species, are heading. As far as we know, we are the first organism to grasp the process of evolution. Other creatures have evolved but we, uniquely, are aware of the process. The religious tendency might be the predictable outcome of a universe with a sense of direction and an evolving species with an imagination.

And God might be, for now at least, the destination, rather than the source, of creation.

5.4 Conclusion

The religious experience of millions, the proven effects of that experience on the human spirit, the drive in man to give life meaning and the similarity in the essential messages of all the world's religions do not, in our view, make the existence of God probable but it is probable that there is some form of reality, some type of consciousness, some mode of existence, to which all the above relate.

That said, the concept of God as omnipotent and omniscient raises insoluble logical problems and cannot be accommodated if we hold to our criteria of fact, reason and probability.

There is also a rational objection to the proposition of a God, whose main purpose was to create life and man, taking 10 billion years to create the former and 13.5 billion to create the latter. Nor is it clear why, if indeed creation was the work of God, He should feel the need to conceal Himself and then make faith in His existence the key to salvation.

We have considered the atheist's view that fear of death is the driving force behind religion. While there may be some truth in this explanation, it is weakened by the inevitable fear of the hereafter which, for the devout sinner, is far more frightening than mere death.

There is another possibility that is worth throwing into the pot. The apparent predisposition for mankind to adhere to religions might indicate, not distant and blurred memories of a past creation, but glimpses of the future, an awareness of where we, as a species, are heading. As far as we know, we are the first organism to grasp the process of evolution. Other creatures have evolved but we, uniquely, are aware of the process itself. The religious tendency might be the predictable outcome of an evolving species with the capacity for understanding the world and the universe in which it lives.

Even if this is so, it leaves open the question of whether the religious experience is a response to some form of external, objective, natural reality. This question cannot be answered definitively on our criteria of fact and reason but, on our probability criteria, given our understanding of the nature of the physical world, the evolutionary process and the sense of direction inherent in both, it seems to us to be rather more likely than not.

And, if it is so, we will probably conclude that the universe is underpinned by a powerful (but not omnipotent or omniscient) creative, benign force, existing outside space/time, which has imbued its creation with a sense of direction and which will fulfil itself in the evolution of a higher form of consciousness which, currently, rather like the peace of God, 'passeth all understanding'.

6. FINAL CONCLUSIONS

1. The universe came into being about 13.5 billion years ago.	See 3.1, 3.2.1 (FACT)
2. Whatever caused the universe to come into being was outside time and space because the Big Bang included the creation of time and space. It was immensely powerful (it generated from a single point – a singularity – a space/time continuum at least 13.5 billion light years across). It included an extraordinary creative capacity (after all, it has produced stars, planets, life and human consciousness).	See 3.1 (FACT)
3. We live in a largely immaterial universe of powerful and invisible forces. That is science, not religion, speaking.	See 3.3 (FACT)
4. The atoms of which matter is composed consist almost entirely of empty space.	See 3.3 (FACT/REASON)

5. The whole of science attests to the ***orderliness*** and ***comprehensibility*** of the universe. Science tells us we live in a largely immaterial universe of powerful invisible forces, governed by intelligible, mathematically definable rules. This does not necessarily mean there was a designer but it does call for an explanation of some sort.	See 3.4 (FACT)
6. It is an inherent feature of our universe that more will come from less. Evolution itself, in physics, in biology and in human consciousness provides abundant evidence of the capacity of existence to manifest emergent properties, properties that transcend the components of the whole and can be explained only as an integral creative function of reality.	See 4.2.1 (FACT/ REASON)

7. The universe has moved from inanimate matter to life; and from life to human consciousness. At each stage, it has taken the raw material of existence and refined and added to it. The inanimate became animate; the animate became conscious. These events require explanation. It is not sufficient simply to take these step changes in the nature of existence for granted or to ascribe them to shakes of the dice. Such complacency is dependent entirely on the hindsight fallacy.	See 3.2.3 and 3.2.4 (FACT/REASON)
8. There is a whole series of settings in the laws of physics which seem to have been finely tuned to allow for the development of an element-rich, stable universe, stable enough for life and human consciousness to evolve.	See 3.2 and 3.2.1 (FACT)

9. Before we jump to the conclusion that the universe must have a design, if not a designer, we should note that, if the universe was not favourable to the development of life and human consciousness, we wouldn't be here to wonder about it. Nevertheless, we do need to consider why, out of all the possible universes, the only one we know exists, operates according to precise rules that enabled the evolution of a mind capable of understanding the rules and wondering why they are so favourable to us.	See 3.2 and 3.2.1 (FACT)

10. The hindsight fallacy consists of taking what has happened and then developing a theory of causation, underpinned, if not wholly dependent, on the fact that things are as they are. The best way to determine whether the hindsight fallacy is being employed is simple. Do not start from what is now and look back. Instead imagine you are in the past looking forward. If you were in the past looking forward, would you think it reasonable to say what happened next was inevitable or at least perfectly probable or, indeed, even possible? The three step changes in existence (the universe, life, human consciousness) were not predictable, probable, possible or even imaginable before they happened.	See 3.2.4 (FACT/REASON)
11. Without recourse to any supernatural explanation, it seems probable that the universe itself has ***a sense of direction***.	See 3.2.3 and 3.2.4 (FACT/REASON/PROBABILITY)
12. We should also consider what might be next step change in the nature of existence. For the first time, man, a product of nature, has the ability to consider the possible future direction of the evolutionary process.	See 3.2.3 and 4.1 (REASON)

13. The current human condition (consisting of human consciousness with self-awareness dependent on an animal body with knowledge of its own mortality) is essentially appalling and ridiculous.	See 4.1 (FACT/REASON)
14. Our chosen way to explore and to attempt to make sense of the human condition is through art. All art forms seem to have one thing in common. They provide an opportunity for selfless consciousness, i.e. consciousness without consciousness of self).	See 4.2 and 4.2.2 (FACT)
15. As in physics, biology and other sciences, so in the arts, we see the ubiquitous phenomenon of emergent properties in which, over and over again, the whole is greater than the sum of its parts. The element that makes the whole greater is immaterial. It is a thought, a sentiment or a design. In the arts, it is a gift of the human mind. It should not be taken for granted.	See 4.2.1 (FACT)

16. The most powerful positive human emotion is love. Love requires the one who loves to give the highest priority to something other than self. Given the human predicament (essentially a limitlessly imaginative and creative conscious mind trapped in a temporal and temporary animal body), the power of selfless love as experienced by the body-bound self is both paradoxical and informative, i.e. it seems the greatest of human joys comes from the subordination or denial of self.	See 4.3 (FACT)
17. The phenomenon of selfless consciousness (as experienced though forms of art, through friendship and love, and through religion) constitutes a type of experience not subject to the usual constraints of time and space. Such experiences are not subject in themselves to any law of organic decay; they are not, in any meaningful sense, located on spatial coordinates; and, while they are certainly temporal in one sense, they often seem to break out from the normal passage of time to touch something timeless.	See 4.2 (FACT/ REASON)

18. Man has a moral sense. Selflessness is a key part of moral consciousness. Again it is paradoxical but informative that individuals, clearly separated from the rest of existence by their bodies, should aspire to selflessness. That suggests that the consummation of morality is love, the love which requires us to subordinate the self to the other, whether that other is God or gods, another individual or humanity as a whole.	See 4.4 (FACT/REASON)
19. The most popular approach to explaining the human condition is religion. Religions are an attempt to answer fundamental questions about existence, to give meaning and add value to human life and to make the prospect of death less burdensome.	See 5.3.5 (FACT)
20. Most religions invite the devotee to have faith in God or gods, representing a creative and generally eternal power. Once again, the message seems to be about attaining a selfless consciousness, requiring the adherent to submit to the will of God and to empathise with or love other people.	See 4.4, 5.2 and 5.3.5 (FACT)

21. People's propensity for faith in God or gods and the similarity in the key messages conveyed by these religions suggests at the very least that humanity aspires towards something greater and better than itself. It remains unclear whether this drive is simply a human aspiration or whether, as religious adherents believe, it is recognition of and communication with an external, objective reality.	See 5.3.5 (FACT/ REASON)
22. If we lean to the view that there is an external reality which we wish to call God, we will probably conclude God is a powerful, creative, benign force, existing outside space/time, but we will have to concede He is neither omnipotent nor omniscient.	See 5.3.2 – 5.3.4 (REASON)
23. Whether or not the religious experience is simply an aspiration or an actual connection with divine reality, it is very probably an excellent indication of humanity's evolutionary direction.	See 5.4 (PROBABILITY)

24. The apparent predisposition for mankind to adhere to religions might indicate glimpses of the future, an indication of where we, as a species, are heading. As far as we know, we are the first organism to grasp the process of evolution. Other creatures have evolved but we, uniquely, are aware of the process itself.	See 5.4 (FACT/ REASON/ PROBABILITY)
25. The religious tendency might be the predictable outcome of a universe with a sense of direction and an evolving species with an imagination. And God might be the destination, not the source.	See 5.3.5 (REASON/ PROBABILITY)

26. Even if this is so, it leaves open the question of whether the religious experience is a response to and communication with some form of external, objective, natural reality. This question cannot be answered definitively on our criteria of fact and reason. But, on our probability criterion, given: a) our understanding of the nature of the physical world (a largely immaterial phenomenon governed by powerful forces that act according to intellectually comprehensible rules) b) the evolutionary process, including the three extraordinary step changes in reality (the creation of the universe, the birth of life and the emergence of human consciousness) and c) the sense of order and direction inherent in both of the above it does seem to us to be, on balance, rather more likely than not.	(PROBABILITY)

27. So, if you accept the facts in this book and you are guided in your thinking by reason, you are compelled to conclude that, while there cannot be certainty, the most likely truth is that there is a generally benign, creative force infusing the universe in which we live.

7. AN AFTERTHOUGHT

One further thought that has no place in a treatise committed only to fact and reason. Think of it, if you like, as a bit of eccentric speculation. But grant that it does answer quite a few, intractable questions. And I hope it will go some way to satisfy those who took the title of this book to be a promise rather than an analytical teaser.

If God is mankind's destination, the next step change for existence (after the creation of the universe, the birth of life and the emergence of human consciousness) could be a form of self-less but individuated enhanced creative consciousness on a plane unconstrained by the constructs of time and space.

Is that too much of a speculative leap? Perhaps but it's not so far-fetched, given that many of us, in living our lives, exploring ideas, enjoying the arts and experiencing love, already have some insight into the self-less consciousness that such a step change would entail. And surely not any more improbable or less conceivable than matter turning into life; or life creating human consciousness?

If that were so, the 'God' who then came into being, wholly positive and possessed of unlimited creativity, exempt from all temporal and spatial considerations, could also be the God of creation, the God whom the prophets glimpsed, the God who was hidden from the rest of us because He had not yet been fully realised. Once fully realised, God would effortlessly encompass eternity (i.e. all the time that ever has been and ever will be). In other words, when He comes into being, He will have always existed and could therefore be the creator of the universe as well as its mentor and its fulfilment.

It is of course a paradox. But it would explain some of life's mysteries. It would explain why the prophets have provided confusing accounts of the nature of God; why, being hidden from us, He is known to mankind only through faith; and why the world and our position in it is a little bit messy. It is simply because, in this, our universe, God has not yet been realised, although, when He is fully realised, since He exists outside time and space, He will, insofar as space and time will then have any meaning, have been around all the time.

How is God to be realised? No surprise there! Through love. Through the development of the self-less consciousness that underpins art, morality and religion, a form of consciousness to which we can all contribute and in and through which the best of each of us can achieve immortality and finally be at one with each other and with God.

And, when this God, our God, is fully realised, He will, for us, be omnipresent. He will then also be able to be wholly good and omnipotent and omniscient without the contradictions that arise for such a God in a temporal world.

Yes. I know it's a stretch. But give it some thought.

There, at last as promised, is God for the curious unbeliever.

(Objection: 26)

8. APPENDIX: OBJECTIONS & ANSWERS

Objection 1: I have faith and faith is more important to me than fact or reason.

Fair enough. This book is not for you; it is for the curious unbeliever.

My problem with faith is that it is inherently undisciplined and unconstrained and, for obvious reasons, precludes rational, critical discussion.

Objection 2: I don't want to think about fundamental questions.

No one will blame you. Such questions are hard and can be disturbing. But it means you have to live your life in ignorance and denial. You will be refusing to engage in what most thinking minds consider to be the most fundamental questions of human existence.

Socrates suggested that 'the unexamined life is not worth living'. It's an uncompromising statement but I think he meant that those who fail to address the types of questions posed here are not participating fully in the experience of what it is to be human. Or perhaps he just meant that life is an unearned gift with unlimited potential; and it's a great shame not to explore it as thoroughly as we can.

Objection 3: There are no facts. You can't be certain of anything.

There may be few facts of which we can be absolutely certain but, in everyone's life, there are facts which we have to accept in order to function. For example, we need to breathe, to drink and to eat, or we die. It makes no sense to reject all facts just because we cannot be certain there are any 'absolute' facts, or just because some facts turn out to be wrong. It is in man's nature to try to understand. Facts are the bits of understanding that we are pretty confident are much more reliable (i.e. correspond with reality) than fantasy or lies.

Objection 4: Facts are changing all the time.

Yes, of course. That's what man does. He learns by experience and reason. He improves his grasp of facts. The important conclusion to be drawn from the replacement of the Ptolemaic view of the solar system by the Copernican view is not that the Ptolemaic view was wrong but that the Copernican view was 'more right' in that it fitted our observations better.

Objection 5: Reason has its place but I don't think it's that important.

There are two sets of people who present this view. There are those who have a serious philosophical objection to classical rationality and there are those who are simply not capable of reasoning.

To the first group, we say that we are using reason as a criterion in its simplest sense, as a means of eliminating inconsistent propositions. It's true that quantum theory seems to be inherently paradoxical but, at the level of day to day matters, reason seems to be the soundest foundation for thought.

The second group (those incapable of reasoning) are more of a problem. Sadly, in recent years, people have been encouraged to ignore inconsistencies in the interests of tolerance or inclusivity. It is the fashion now to argue that people are entitled to hold entirely contradictory views so long as they feel comfortable with them. If you are someone who feels that you can disregard reason, if you are not challenged intellectually by inconsistencies in your own argument, if you are not critical of inconsistencies in other people's arguments, if, in short, you disrespect reason, then this book is not for you.

There are others in this group who take some pride in their rejection of reason. In the 11th century, Al Ghazzali, one of the most revered Muslim scholars, repudiated reason and, in his rejection of *falsafa* (philosophy), was a key figure in turning Islam away from logic, mathematics and physics, on the grounds that they were incompatible with Islam. The consequences of that rejection of reason are cruelly etched in the history of Muslim countries from the 12th century on.

Today, in many religions, there are those who ignore facts and decry reason as enemies of faith. The views of such people don't make much sense in that, if they believe in God, then surely they need to understand His Creation, its character and purpose. Their views seems irrational – but then, since they reject reason, such a criticism is unlikely to cut much ice with them. So it's probably best we say goodbye to them – and all who reject reason – at this point.

Objection 6: If you can't be certain you have the absolute truth, and no one can be, you might as well not even try to explore the fundamental existential questions.

I have some sympathy with this view. We all want certainty and it's frustrating that, unless you have faith, you are most unlikely to find (or believe you have found) absolute truth. The history of mankind has been one of testing, checking, rejecting, refining. What we think now is not absolutely true but it is truer than what we thought before. There are not four elements; there are 118. The Sun does not go round the Earth; the Earth goes round the Sun. Diseases are not caused by noxious vapours but by bacteria and viruses. The truth is work in progress. But surely that means we should work on it, not quit?

Objection 7: I accept facts and reason as bases for exploring fundamental questions – but probability? Surely using probability takes us into the realms of faith?

Again, I see the argument but, unless we accept the probability criterion, we will find our way blocked almost immediately. Unless we expect to find absolute truth, we are going to be relying on some degree of probability from the start. But we will not use the probability criterion indiscriminately. We will use probability only when we cannot resolve a question by reference to the facts or by the application of reason. In such circumstances, we justify probability simply on the grounds that, if you seek to resolve a question and you can envisage several possible answers, it seems reasonable to opt for the one that is most probable, i.e. the one that, based on our experience and understanding of life, tested by reference to the facts and the application of reason, seems most likely to be true.

Remember, we are not expecting to find 'the Truth'. We are merely trying to get as close to the truth as our current knowledge of the facts and the application of reason will allow.

Objection 8: 'Why is there something rather than nothing?' This is not a real question. There never was 'nothing'.

OK. Not 'Why is there something rather than nothing?' but 'Why is there something?' There is definitely something. Surely that needs explaining. Even if that something had always been there, it would need explaining but, given that, according to our current understanding of the facts, the universe, *including time and space*, came into being 13.5 billion years ago, the necessity to ask the question is undeniable.

Some people have a real problem with this question. They take existence for granted; they simply don't understand the question. The irony of such a viewpoint is that almost everything such people believe to be self-evidently real, so real its existence (its quality and substance) cannot be questioned, turns out to illusory. (See 9, below.)

Objection 9: I really don't see the problem. The universe exists. It is given because it's there.

I understand this objection to the question. It is the 'common sense' approach. Just accept what you see with your eyes, hear with your ears, etc. It's tempting but, as it turns out, completely irrational because it is inconsistent with the facts.

The senses are profoundly misleading if we want to grasp the nature of reality. Our senses give us access to an incredibly narrow range of information about reality, and most of what we conclude about reality from our sense perceptions turns out to be wrong. The best description of reality, based on facts and reason is this: we live in a largely immaterial universe of powerful and invisible forces, governed by intellectually coherent rules.

That's not exactly what the advocate of 'common sense' or the conventional materialist wants to hear. But, according to science, that's about as good a description as we can formulate. Of course anyone is free to cling to the 'common sense' view but they must concede they do so as an act of faith and against the known facts and reason.

Objection 10: Why do you put atheists in the same category as those who believe in God? It is rational not to believe in the existence of an invisible, unprovable entity.

If you accept the arguments set out in answer to the question 'Why is there something?' you need faith to argue that everything 'just is', i.e. uncaused and given. True, the universe exists and the atheist argues that God does not exist but belief in an uncaused universe requires faith, just as belief in God requires faith. The atheist simply nominates the universe, rather than God, as the First Cause. Belief in an effect without a cause requires an act of faith.

The only position consistent with the facts and reason is that the universe came into being as the result of a powerful creative force (character unknown) outside time and space.

Objection 11: Cause and effect are merely mental constructs. There's no reason why a phenomenon must have a cause.

Apart from the paradox of using reason to deny the necessity for a rational explanation, this position is perverse. Cause/effect relations are not merely mental constructs; they are observed in all natural phenomena. And, given that we observe cause and effect in all natural phenomena and find intellectual satisfaction in discerning cause/effect relations, it would seem perverse (i.e. against the facts, reason and probability) to deny the need for an explanation of the totality of existence.

Objection 12: I don't see what is so 'particular' about our universe. It is what it is. Isn't it like asking why water should be wet?

The universe is governed by rules. The rules determine what there is and how it behaves. These rules, in many cases, seem to have been calibrated to create not merely this universe, but this particular universe. If the rules were only slightly different, there might still be a universe but it would be a very different one from the universe we inhabit. Some of these rules, like the strength of gravity, seem arbitrary but, if they were not as they are, our universe and we would not exist.

Don't take things for granted. The curious mind would like to know why there are rules and why, seemingly against all the odds, the rules appear to be favourably disposed to the emergence of life and human consciousness.

Objection 13: But surely any universe that existed would have to be finely calibrated, i.e. it would have to be the result of specific rules.

Not really. There's no obvious reason why a universe should be governed by any rules. In fact, it seems more likely that a random emergence of existence would be chaotic rather than ordered. The fact that our universe is intelligible (i.e. is sufficiently ordered for the rational mind to be able to comprehend it) is in itself deeply thought provoking.

And even if there are those who take the existence of an ordered universe for granted, and then argue that, if the universe is ordered, then obviously it has to be governed by rules (two extraordinary leaps of faith, by the way), they must still explain why the specific rules that govern our universe are what they are. Intelligent life has emerged on at least one planet. If the rules were not as they are, were not the particular rules of our universe, life would not have been possible, much less intelligent life. That is one of the reasons why scientists have postulated the multiverse theory which envisages the existence of innumerable other universe where the rules are differently calibrated and where, presumably, life has not emerged. The *particularity* of our universe needs to be explained and the significance of the particularity needs to inform our views on the nature of existence.

Objection 14: You suggest the universe has a sense of direction. What does that mean?

It's a good question. Currently the received wisdom is that over billions of years primordial gases (predominantly hydrogen and helium) coalesced into stars. Exploding stars created planets, probably trillions of them. On at least one planet, eventually life emerged. And after another few billions years, human consciousness evolved. All this is derived from the facts as we know them.

You can argue that all this happened by chance and certainly chance is a possible explanation. But even the brief account of the history of the universe given in the preceding paragraph seems to me to suggest an alternative and more probable explanation. There have been three step changes in the nature of existence. First out of nothing, something comes; then, after billions of years, out of inert matter life emerges; then a few billion years after that, human consciousness evolved. Yes, it's an inordinately slow process but surely we see a trend: the universe is becoming conscious of itself.

And, if they had not happened, each of the three step-change events in existence would have been unbelievable. No rational mind would have predicted that an entire universe would emerge out of miniscule singularity; that matter, inert for almost 10 billion years, would find a way to become animate; that life, driven by the desire to survive and reproduce, should transcend the immediate demands of fighting, fleeing and copulating to evolve a mind capable of extraordinary understanding and creativity.

Yes, it could all have happened by chance but given the three existential step changes, all pointing in the same direction, it seems to me more probable that there is a discernible trend in the evolution of the universe or, as I like to put it, the universe has a sense of direction.

Objection 15: Aren't you yourself guilty of the hindsight fallacy by finding a trend, looking back, and then imposing your own preconceptions on the data?

No. Those who employ the hindsight fallacy are explaining away improbable events by looking back and postulating possibly spurious cause and effect patterns.

In a way, I'm doing the opposite. I'm pointing out that the step changes in the nature of existence were unforeseeable and are so extraordinary that, given the infinite number of possible scenarios in which these step changes would not have occurred, they call for a deeper explanation than chance.

I'm also suggesting that, when taken together, the step changes strongly suggest a trend.

Objection 16: I'm not happy with the concept of a sense of direction. The evolution of life doesn't have a sense of direction. It works simply on the basis of random mutations and the survival of the fittest.

You're right to question my 'sense of direction' hypothesis. For me, it's a crucial concept.

It seems to me that the theory of evolution itself involves a sense of direction. Taking as given for a moment the first two step changes in existence (the universe itself and the emergence of life), it is reasonable to ask why the life that emerged was imbued with life's particular characteristics. Why was it determined to survive? Why was it driven to replicate and reproduce? Why didn't it realise it was a freak accident, born into a profoundly hostile environment, and quit? The survival instinct explains why we want to survive; it doesn't explain why we have such an instinct.

If you follow the evolution of species, again it is arguable that life has a sense of direction.

> *'O flightless, feathered reptile, strive*
> *for aeons just to stay alive,*
> *not knowing those long finger things*
> *will in the end support your wings.'*

> from the poem *Given Time*
> by the author

It's not unreasonable to wonder why species are prepared to accept obvious short-term disadvantages in order to achieve long-term benefits. Did the earthbound dinosaur look with envy at the vast three dimensional space afforded by the sky and think 'One day...?' Or is a drive to explore and eventually become conscious of the universe somehow inherent in creation.

None of the above is intended to prove that the universe is the creation of a designer. The existence of a designer is a possible but not necessary explanation. The hypothesis that the universe has a sense of direction is simply a conclusion that seems to fit best with the facts.

Objection 17: It's possible that the universe has a sense of direction but I prefer to believe the universe and we are just a truly tragic and/or ironic accident

That's a point of view and no one can conclusively prove such a theory is wrong but, of all the alternative explanations of existence, is it the most probable?

How well does it account for the existence of the universe, for the birth of life, for the emergence of human consciousness? Is it just an accident that the universe is governed by rules, that it performs according the mathematical formulae, that it is comprehensible to the mind of man? Does the theory explain the widespread drive in man to seek meaning and other similar transcendental aspirations? How well does it account for what seems to be man's innate moral sense? Does it explain man's creativity?

If another theory seems to answer some of these questions more persuasively than the tragic/ironic accident hypothesis, is it not reasonable to decide this other theory is more probable?

Objection 18: You say we live in a largely immaterial universe of powerful and invisible forces, governed by intellectually comprehensible rules. That's not my world.

Of course that is not the day to day world in which we live our lives. In our lives, we are compelled to view the world through our senses. We are constrained by our individual bodies, and evolution has ensured we, as individual organism, are finely tuned to respond to and deal with our environment.

The essential point is this. The world we experience through our senses is not the simple, straightforward, rock solid reality we think it to be. Those who believe it is are deceiving themselves. I know it's hard but, if you pride yourself on the common sense approach of restricting your ideas to what you can be certain of, you need to open your mind to exactly what that is. And what is it? Well, in brief, it is a largely immaterial universe governed by powerful and invisible forces which perform according to intellectually comprehensible rules. So, like it or not, the senses are not a terribly good guide to reality.

Evolution has given us a mind and that mind gives us the opportunity to go beyond the constraints of our senses. Through the arts and sciences we can explore reality. We can now contribute to the process of the universe becoming conscious of itself. We do not know where the process will end, any more than inert matter foresaw life, or primitive life foresaw the mind of Mozart or Einstein but, in general terms, we can see where, in all probability, we are heading.

Objection 19: What's so remarkable about 'emergent properties'?

Well, it's odd that the whole can be greater than the sum of its parts because it means that whatever makes the whole greater is something immaterial. It prompts the thought that, although our senses and much of our day to day experience emphasizes the materiality of the world, almost everything that really matters (love, friendship, loyalty, appreciation of art and, for those with faith, religion) are essentially immaterial.

I suppose the most familiar outstanding example of the whole being greater than the sum of its parts is the human brain. The fact that 1.4kg of grey matter was capable of producing Michael Angelo's *David*, Beethoven's *Ninth Symphony*, Newtonian physics, the theories of evolution and relativity and quantum mechanics must surely give everyone pause for thought.

At a rather less exalted level, we might even express wonder at the way in which meaning emerges from the combination of words in a sentence, surely the simplest and most ubiquitous example of the whole being greater than the sum of its parts.

We are so used to the whole exceeding the sum of its parts that we take it for granted but we shouldn't. It is an amazing feature of existence.

Objection 20: I don't know what you mean by self-less consciousness. Sounds like psycho-babble to me.

There are different types of experience. If I burn myself, my self is very much at the centre of my consciousness. It is my hand that is burnt, that hurts, and that is the only thing I am thinking about. If my friend burns his hand, I feel sympathy for his pain. To some extent, I feel his pain; but my self is not at the centre of the experience. If I love someone, it is still my self that is loving but I am happy to put that person before my self. When I listen to the 'Halleluiah Chorus' or Beethoven's *Ninth Symphony*, I would say I am experiencing heightened consciousness but my self is scarcely present.

In other words, there is a spectrum of human consciousness which runs from almost totally self-centred to almost totally self-less consciousness. Admittedly, this assertion is empirically-based, rather than a product of fact or reason, but in this section we are simply exploring human experience to look for pointers in drawing conclusions about our true nature – and most people will have experienced at moments in their lives some degree of self-less consciousness.

Objection 21: You say that it seems that the greatest of human joys comes from the subordination or denial of self. Really? What about the pleasures of the flesh – sex, eating, drinking, etc.?

No one would deny sensual pleasures. We have already conceded we are half animal, subject to the bodily drives that ensure the survival of the species. To these, we have added a few from which the animal world is largely precluded, drives such as avarice, schadenfreude, sadomasochism.

That said, most would agree that the greatest of human joys comes from love, whether secular or sacred, whether of the erotic, platonic or spiritual type.

Objection 22: Of course people may vary according to various criteria but, at the end of the day, we are all equal.

If you concede that, according to all the yardsticks of value, individuals vary widely in their attainment level and worth, it is difficult to see how it is reasonable to conclude, "at the end of the day", (i.e. when all these qualities are taken into account) that all individuals can somehow be assessed as of equal value.

It is possible to assert, as a basic premise, that all individuals are of equal worth; but it is a premise, not a logical deduction. And, at least to some extent, such a premise is irrational. The concept of worth involves a value judgement and, as we have argued, by any criteria of value, it is obvious the attainments of individuals vary widely.

This analysis is not intended in any way to deny or denigrate a moral sense; it is simply intended to point out that reason is a shaky foundation on which to base it.

Objection 23: Yes, man has a moral sense. But that is perfectly explicable in terms of social demands, evolutionary self-interest, etc. There's no need to bring in religion.

Quite right. There are several possible causes for man's moral sense. No one is arguing a moral sense is necessarily a sign of human spirituality. We draw only two points from this section. First, that man has a moral sense and that fact requires some explanation. Secondly, all forms of morality seem to demand that the individual denies certain 'base' selfish desires in favour of 'nobler' selfless sentiments. Of course the words 'base' and 'nobler' are value-loaded words but the distinction between selfishness and selflessness is objectively observable in the behaviour of those who espouse most, if not all, moral codes.

There is of course an ambiguity, if not a contradiction, at the heart of the concept of self-less consciousness. Many would suggest that the self is what gives individuals identity. How can there be human consciousness of a high order without the self? And yet, in various types of heightened human experience, that seems to be what happens. In love, in religious ecstasy, in profound appreciation of the arts, even sometimes in the acts of creation, the self seems to be suppressed, irrelevant or absent.

Objection 24: The driving force behind most religions has been the desire of the priestly class, or those whom they served, to exercise control over the masses.

There has certainly been an element of *power over the people* in most religions as they have developed but, as far as we can tell from the lives of the prophets, the origin of their revelations was a sincere belief that they had accessed some level of truth which they were then compelled to promulgate.

Of course as soon as they or their followers became organized, they developed political and social structures which they then used to acquire and hold on to power.

It is also true that the success of many religions has been achieved and sustained through the ruthless exploitation of people's wishful thinking and fear of death.

Nevertheless, the similarity of religious revelations, the sincerity of the prophets and the religious experiences of millions should not be lightly dismissed. And that's all I am arguing.

Objection 25: Throughout, you have persisted in equating atheists with believers. That's not fair. There is a difference between those who believe in the unprovable and those who don't. The former are fantasists; the latter are rational.

When we began this journey, we agreed that there were fundamental questions which most of us at some time ask ourselves. We agreed our purpose was to apply three criteria in our pursuit of answers to these fundamental questions. The questions are difficult and we accepted we probably wouldn't find incontrovertibly true answers but we agreed to do our best.

Militant atheists refuse to answer any of the key questions. Their answer is there is no answer. I find that irritating. The questions we have been considering are the most profound and challenging any human being can ask. Why is there something rather than nothing? Why is there this something? Why should this something be subject to physical laws that are intellectually comprehensible? Why should inert matter remain inert for 10 billion years? Why did it then become animated? What set the direction for hydrogen and helium to generate 'Beethoven's Ninth'? What made that transcendental outcome possible? Evolution is a mechanism to explain how things happened. It doesn't explain why there was an evolutionary mechanism. What is going on? How is it possible for the whole to be greater than the sum of its parts? Come on! No one has the answers for sure. But for the rational mind, to say there are no answers is surely as stupid as saying God created the world in six days.

Let's say that your love for someone is a circle. You want to quantify your love (i.e. measure the area of that circle).

I say: 'Multiply the radius by itself and multiply the result by 22/7. You won't get a precise result but you will be close.'

The theist says: 'Good try, but you will never know the true answer through human calculations'.

The agnostic says: 'Interesting but sadly the result is inconclusive and imprecise'.

The atheist says: 'There's no such thing as love'.

(Note from the author to the author: don't ever use analogies.)

Objection 26: What you suggest in your Afterthought, is paradoxical nonsense. If we are part of the process of creating God, surely God doesn't currently exist, does He/She?	Yes and No.
In any case, do you have any firm evidence to support such extraordinary speculations?	No.
Are we really to take this afterthought seriously?	Yes.

www.ingramcontent.com/pod-product-compliance
Lightning Source LLC
Chambersburg PA
CBHW050541300426
44113CB00012B/2218